baby FROGS

KIM THOMPSON

CREATIVE EDUCATION • CREATIVE PAPERBACKS

CONT

ENTS

I AM A FROGLET.

I am a baby frog.

See my tail?
It will go
away soon.

In the spring, my mom laid hundreds of eggs in a pond. They looked like jelly.

When I hatched, I was a tadpole.

I swam and ate plants. My legs started to grow. Lungs formed. Skin grew over my gills.

These big changes are called a metamorphosis.

It is summer.
My tail goes away.

I can live on land. I can catch insects. Finally, I am a frog.

Now I can jump!

SPEAK AND LISTEN

BBIT!

Can you speak like a frog?

Little frogs whistle and call.

Listen to these sounds:

https://www.youtube.com/watch?v=lllzh1rJyPk

Now it is your turn!

FROG WORDS

gills: body parts used for breathing underwater

lungs: body parts that help animals live on land and breathe air

metamorphosis: a series of big changes that frogs, butterflies, and other animals go through as they become adults

tadpole: a frog at the larval stage of life; a young frog with a tail and gills

READING CORNER

Kenney, Karen Latchana. *Life Cycle of a Frog.* Minneapolis, Minn.: Jump!, Inc., 2018.

Nolan, Kate. *Usborne Minis: Pond Life to Spot.* London: Usborne, 2021.

Prince, Liz. *Science Comics: Frogs, Awesome Amphibians.* New York: Macmillan, 2023.

INDEX

PUBLISHED BY CREATIVE EDUCATION AND CREATIVE PAPERBACKS
P.O. Box 227, Mankato, Minnesota 56002
Creative Education and Creative Paperbacks are imprints of The Creative Company
www.thecreativecompany.us

LIBRARY OF CONGRESS CATALOGING-IN-PUBLICATION DATA
Names: Thompson, Kim, 1970- author
Title: Baby frogs / Kim Thompson.
Description: Mankato, Minnesota : Creative Education and Creative Paperbacks, [2026] | Series: Starting out | Includes bibliographical references and index. | Audience term: juvenile | Audience: Ages 4-7
Creative Education and Creative Paperbacks | Audience: Grades K-1
Creative Education and Creative Paperbacks | Summary: "Introduce beginning readers to the world of baby frogs with this life science starter. Includes photos, a labeled animal diagram, "Make a Noise" section, glossary, and further resources"-- Provided by publisher.
Identifiers: LCCN 2024043249 (print) | LCCN 2024043250 (ebook) | ISBN 9798889897460 library binding | ISBN 9781682778326 paperback | ISBN 9798889897590 ebook
Subjects: LCSH: Frogs--Infancy--Juvenile literature | Tadpoles--Juvenile literature
Classification: LCC QL668.E2 T48 2026 (print) | LCC QL668.E2 (ebook) | DDC 597.813/92--dc23/eng/20250107
LC record available at https://lccn.loc.gov/2024043249
LC ebook record available at https://lccn.loc.gov/2024043250

DESIGN AND PRODUCTION
Design by Rhea Magaro
Production by Beeline Media and Design, Inc.
Art direction by Tom Morgan

PHOTOGRAPHS by Alamy Stock Photo/blickwinkel/H. Duty, 10-11; Dreamstime/Gustavo Andrade, cover; Shutterstock/Bohbeh, 13, CanuckStock, 14, Dr Morley Read, 9, Joshua Ouellette, 8, Kurit afshen, 2, 4, 12, Marco Maggesi, 6, 7, Mirror-Images, 11, Oleksandr Lytvynenko, 5, Piyapong pc, 7

Printed in India